A Few of My Favorite Things

Counting to Twenty in English and Spanish

Written by Jayne Inoa
Illustrated by Melissa Morris Inoa

JayVee Learning
An INOAVATION Project
www.inoavation.com

Book design by Melissa Morris Inoa

Inoavation Books

A Few of My Favorite Things (Counting to Twenty in English and Spanish)
ISBN 978-1-962953-01-6

For My Diane

J.I.

For Miss Donna and Miss Carol

M.M.I.

one

1

uno

one chocolate bar

una barra de chocolate

two

2

dos

two unicorns

dos unicornios

three

3

tres

three umbrellas

tres sombrillas

four

cuatro

four pieces of jewelry

cuatro piezas de joyería

five

5

cinco

five rainbows

cinco arcoiris

six

6

seis

six flamingos

seis flamencos

seven

7

siete

seven books

siete libros

eight

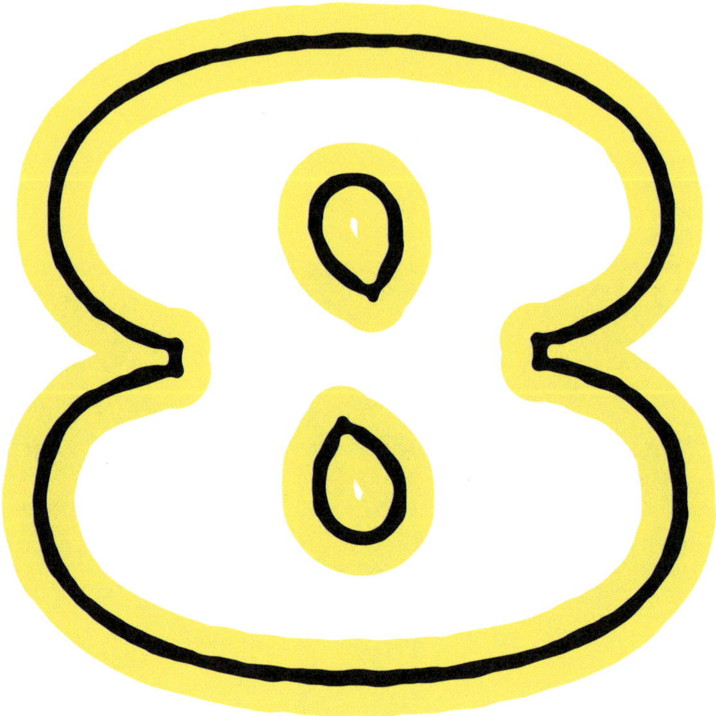

ocho

eight pterodactyls

ocho pterodáctilos

nine

9

nueve

nine cookies

nueve galletas

ten

10

diez

ten dolls

diez muñecas

eleven

11

once

eleven ice cream cones

once conos de helado

twelve

12

doce

twelve flowers

doce flores

thirteen

13

trece

thirteen hair bows

trece lazos para el cabello

fourteen

14

catorce

fourteen nail polishes

catorce esmaltes de uñas

fifteen

15

quince

fifteen bandages

quince curitas

sixteen

16

dieciséis

sixteen leaves

dieciséis hojas

seventeen

17

diecisiete

seventeen wands

diecisiete varitas

eighteen

18

dieciocho

eighteen rocks

dieciocho rocas

nineteen

19

diecinueve

nineteen hearts

diecinueve corazones

twenty

20

veinte

twenty cupcakes

veinte pastelitos

Jayne Inoa

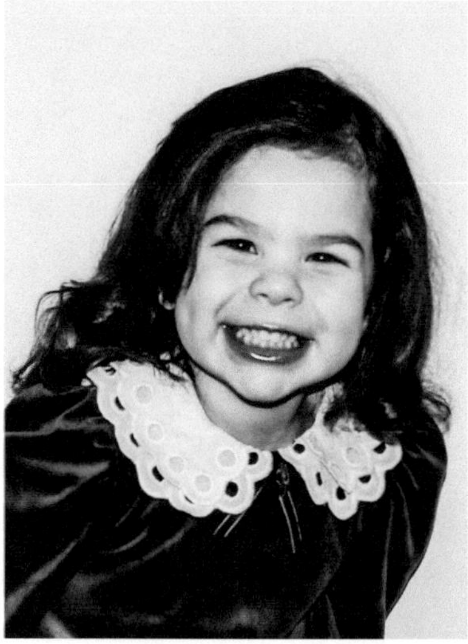

Jayne Inoa is a young writer from New Jersey. Her favorite things to write are song lyrics and (very) short stories.

Besides writing, Jayne loves to sing, run, dance, ride her tricycle, play with friends, and is taking drawing lessons from her big brother.

This is Jayne's first book and was completed when she was three years old.

Melissa Morris Inoa

Melissa Morris Inoa is a writer and educator from central New Jersey. She believes that it is never too early (or too late) for a writer to begin developing the craft and is passionate about helping writers of all ages find their unique voice.

This is the first book Melissa has illustrated. She is looking forward to other projects with Jayne in the future.

Made in the USA
Monee, IL
05 November 2026

67378696R00029